FOR YUMIKO

Hunter and his Dog

BRIAN WILDSMITH

Oxford University Press
OXFORD TORONTO MELBOURNE

On a farm
deep in the countryside
there lived a dog.
One day the dog had
three puppies.

A hunter came to visit the farm.
He bought one of the puppies
to train as a hunting dog.

As the puppy grew into a dog
the hunter taught him tricks.

He threw a stick and made the dog
bring it back to him.

Then the dog learnt how to carry eggs
in his mouth without breaking them.
"Soon," said the hunter to his dog,

"you will be ready to come hunting with me."

One day,
early in
the morning,
they went out
together to hunt.
They waited
in the reeds.
Suddenly
a wild duck flew up
into the air.
The hunter
raised his gun.
Boom!
Slowly the duck
fell to the ground,
wounded.

Off went the dog to bring back the duck.
When he saw it, hurt and helpless
on the ground, he was sad.
Gently, he picked up the duck in his mouth
and carried it to a little island.

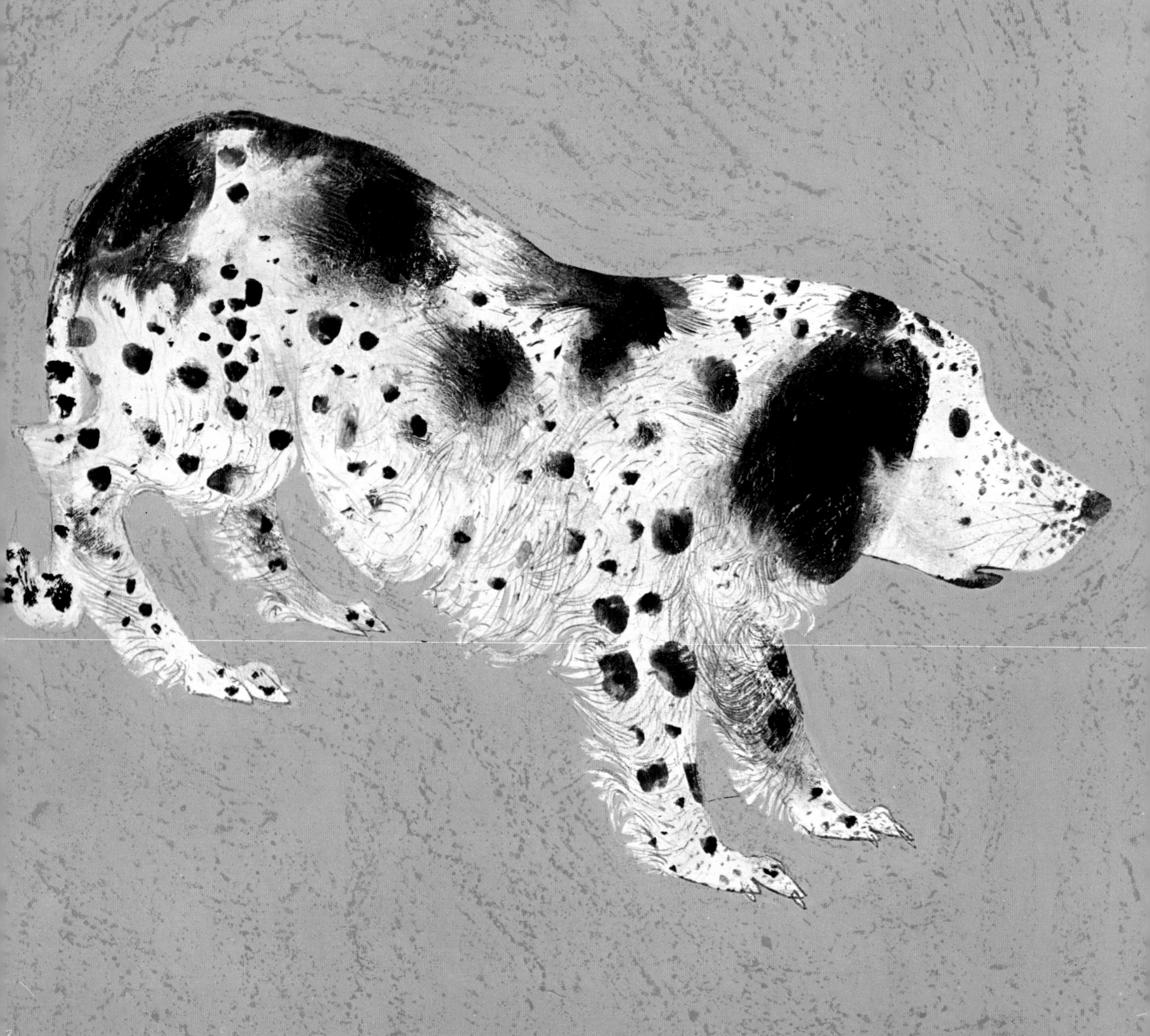

The dog licked the duck's wounds.
"I will look after you," he seemed to say.
Then he played a trick on the hunter. "I was taught to bring back a stick," said the dog to himself,

“and so a stick I shall take back to him.”
So he left the duck where it was,
and returned to the hunter with a stick.

Again they went hunting. Every time the hunter shot at a duck, the dog went after it. Every time

the dog found a wounded duck he took it to the
island and came back with a stick for the hunter.

And every night the dog stole a loaf of bread from

the hunter's home and went off to feed the ducks.

One night the hunter saw
the dog take the bread.
He followed the dog outside
and saw him go into the reeds.

The hunter
peered through
the reeds.
He saw the dog
feeding
the ducks
with bread.
The hunter was
full of shame
for what
he had done.

In the morning the hunter brought
a cage to the island.

Tenderly he put the wounded ducks into it
and took them back to his home.

He bathed the ducks' wounds

and bandaged them to make them better.

When the ducks were well again, he took them out one morning to the place where he had shot them.

One by one he let them fly away
into the rising sun.

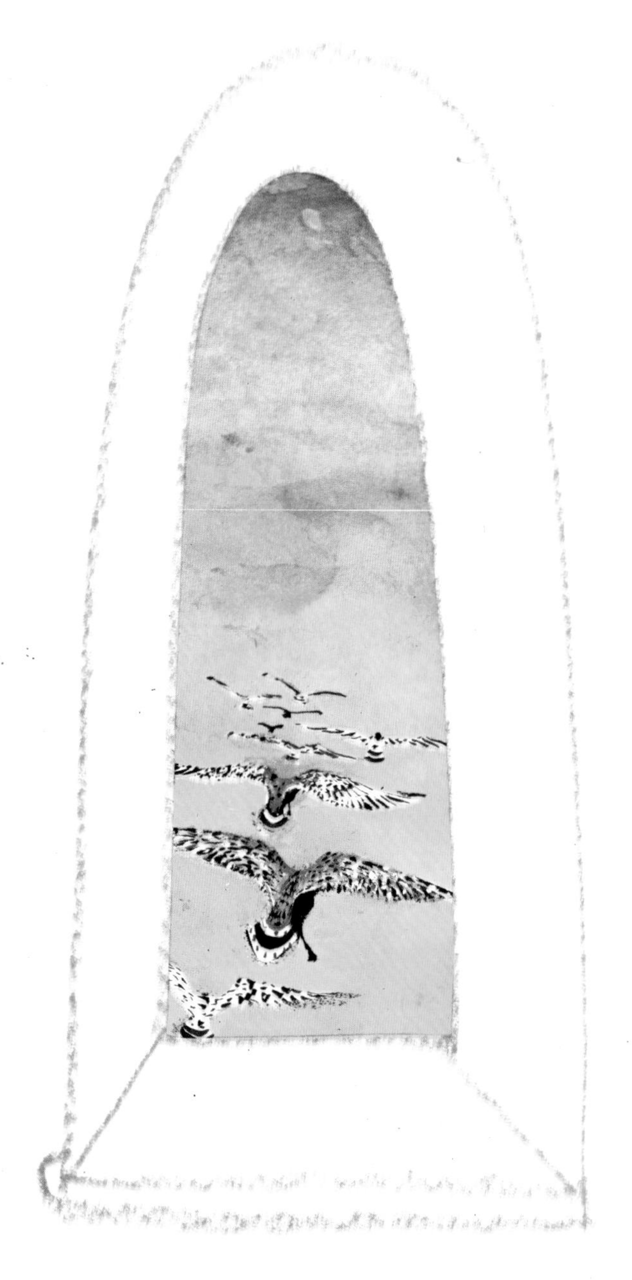

Oxford University Press, Walton Street, Oxford OX2 6DP

OXFORD LONDON
NEW YORK TORONTO MELBOURNE AUCKLAND
KUALA LUMPUR SINGAPORE HONG KONG TOKYO
DELHI BOMBAY CALCUTTA MADRAS KARACHI
NAIROBI DAR ES SALAAM CAPE TOWN

and associated companies in
BEIRUT BERLIN IBADAN MEXICO CITY NICOSIA

Oxford is a trade mark of Oxford University Press

ISBN 0 19 272147 X

First printed 1979
Reprinted 1980, 1984
First Published as Paperback 1984
Reprinted 1985

PRINTED IN HONG KONG